Who Eats What? EDITION

Ruth Komesaroff

A word from our editor

WE all need food to stay alive. It is the fuel that keeps our bodies working and growing. When we are hungry, we can reach for a snack or look for food in the fridge.

But have you ever wondered how all the animals, birds and fish in the world get their food fuel? What do they eat? How do they find their food or catch it? How do they hide from other animals who are trying to eat them?

Read this edition of *The News*, and you will find out who is eating what.

Daily spot

Some spiders and insects kill and eat their partners after mating. The female black widow spider and the female praying mantis are larger than their mates. After mating, the female kills her partner.

The chain of life

Links in the food chain.

EVERY LIVING thing needs food. Food provides the energy that keeps everything alive and growing. Imagine a chain. Animals and plants are part of this chain because of what they eat. The gazelles eat grass, the cheetahs eat the gazelles, and sometimes they are both eaten by lions.

Each animal is a link in the **food chain**. Each link in the chain gets energy from the links before.

The food chain starts with the sun. It gives off light, heat and energy. The first link in the food chain is plants. They make their own food by using the energy from the sun.

The next link is animals who eat plants. They might eat the whole plant or different parts of the plant—like the fruit, seeds, leaves, bark and roots. Animals who eat only plants are called **herbivores** (say *her-biv-ors*). These animals get their energy from plants.

Gazelles are herbivores. They live on open plains and eat grasses, seeds, shrubs and new green shoots. They are eaten by cheetahs, lions, leopards and hyenas.

The next link in the chain is the animals who eat meat. They are called **carnivores** (say *car-niv-ors*). To get meat in the wild, they have to kill other animals.

Animals who kill other animals are called **predators** (say *pred-ah-ters*). The animals predators hunt are called **prey** (say *pray*).

Other animals in the food chain eat both plants and animals. These non-fussy eaters are called **omnivores** (say *omm-niv-ors*).

Some predators eat other predators, as well as eating herbivores. For any animal in the wild, it is a race to find enough food to survive and to escape being eaten.

Predators can be as large as a lion or as small as a ladybird. Frogs feed on worms and insects. Ladybirds feed on insects that are harmful to plants, such as aphids and scale insects.

Cheetahs are carnivores. They eat gazelles, wildebeests, zebras, hares and birds. Cheetahs are eaten by eagles, humans, hyenas and lions.

Most humans eat a huge range of plants and animals. Humans are part of many different food chains.

Speed freak

Cheetah in high-speed chase with gazelle.

T HE CHEETAH uses speed to stay near the top of the food chain. There was a high-speed chase today on the plains in Africa. The world's fastest land animal, the cheetah, was speeding while chasing a gazelle. The cheetah hit 110 kilometres (68 miles) per hour and reached its top speed in just three seconds.

The gazelle was not quite fast enough, clocking up a speed of just 95 kilometres (59 miles) per hour. The chase was over in 30 seconds.

The cheetah's long legs and springy spine helped it reach its super speed. It also used some fancy footwear. Its claws stick out and act like spikes in sprinting shoes.

The cheetah used a second set of razor-sharp claws to knock the gazelle to the ground. It killed the gazelle by biting its throat.

But the cheetah paid a high price for its fast food meal. It nearly died of heat stroke as its body temperature rose to a dangerously high level.

A gazelle is part of the cheetah's food chain, so every day is a race to survive, while finding enough to eat for itself.

Night-time hunters

Scorpion is part of its own food chain.

Scorpions and hedgehogs love to eat the same kind of food. But they better not eat together or the scorpion will end up on the menu, too!

D URING THE day in the desert, the temperature of the sand can reach a scorching 70°C (158°F).

The sand scorpion and the desert hedgehog beat the heat by sleeping in the daytime. After the sun sets, these predators come out of their hiding places to hunt down their dinner.

These two creatures have similar places in the food chain when it comes to eating out in the desert.

They both enjoy a hearty feed of insects and spiders. An evening meal could also include beetles, termites, crickets and ants.

A spicy treat that they both love is scorpions! The desert hedgehog prepares this dish by nipping off the poisonous stings before eating up the scorpion.

The sand scorpion will even eat its own relatives. It will snack on sand scorpions as well as other **species** of scorpion.

The best known hedgehog is the European hedgehog, which is famous from fairytales and fables, such as Aesop's "The Fox and the Hedgehog".

Did you know?

Scorpions can go for very long periods without food. One species of scorpion has survived for two years without eating any food at all.

Reptiles in the rainforest

Sharp teeth and darting tongues make snacks simple.

RAINFOREST reptiles are clever at maintaining their place in the food chain.

Spectacled caimans (say *kay-muns*) lurk in murky streams and on river banks in South America, waiting for a tasty treat. They eat crabs, fish—including piranhas (say *purr-ar-nuz*)—frogs, snakes, water birds and wild pigs.

Caimans are very fast swimmers. They have strong jaws and long pointy teeth. One of their tricks is to grab their prey on land and drag it into the water. Then they hold it underwater until it drowns.

The chameleon (say *cam-eel-ee-on*) is a small lizard that lives in trees. It's known for being able to make its skin turn green, yellow or brown.

Chameleons have very long, sticky tongues. Their tongues are longer than their bodies, and can shoot out faster than they could ever move themselves. They use their tongues to snap up insects and spiders.

To help them find their food, chameleons have eyes that can swivel around. Each eye can look in a different direction at the same time.

A chameleon's long tongue helps this lizard to keep its place in the food chain. It is the length of the body and the tail combined. Its tongue shoots out so fast that food fuel such as crickets can't get away.

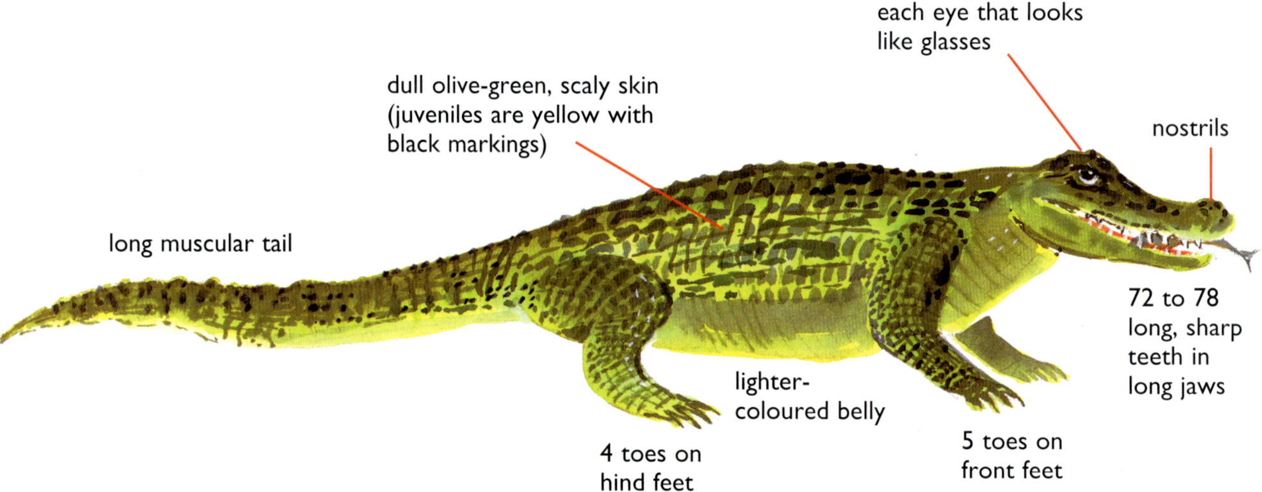

Here is a guide to help you identify a spectacled caiman—so you don't become part of its food chain.

On the alert

Fierce fish is also a tasty dinner.

THE PIRANHA is a vicious predator. These fierce fish are always on the alert for more food fuel.

They eat fast and they don't chew. As an ominvore, they'll eat anything they can sense and catch. Small fish and some larger animals are devoured in seconds. Fruit and nuts which fall into the water are also part of the piranha's diet.

They have special eyesight so they can see a meal in pitch-black water. The water looks black to us, but for the piranhas, it is alive with food.

Piranhas use ripples in the water to work out where their next meal is coming from. They have a sensing device on the side of their bodies which picks up ripples from a passing prey, telling them where their victim is.

But even a fierce predator such as the piranha isn't safe. The caiman is higher on the food chain. It enjoys a piranha or two or three for breakfast, lunch and dinner.

Many piranhas have blunt noses and lower jaws that stick out farther than their upper jaws, making them look a little like bulldogs. The largest piranhas grow to be about 60 cm (2 ft) in length.

Animal of the day

PIRANHA—a carnivorous fish that lives in South American rivers. It has razor-sharp teeth and strong jaws. It swims in large groups with other piranhas. These groups are called shoals.

Amazing adventures

Imagine… what you would see if you spent five fabulous days on safari in the grassy plains of Africa watching food chains at work.

Monday

We saw a herd of elephants this morning, grazing in the grasslands. Suddenly, they started stomping their feet on the ground. Our guide told us that elephants do this to warn each other of danger. Their thumping can be felt by other elephants 50 km (31 miles) away. The distant elephants pick up the ground vibrations through their huge feet.

Leopard

African wild dog

Tuesday

Today we came across a pack of wild dogs. They made a circle around a zebra, then one of them grabbed it by the lip. The zebra went very still and quiet. By biting its lip the dog sent the zebra into a zombie-like state. Then the other wild dogs went in for the kill.

Wednesday

A very strange thing happened today. I saw monkeys fall out of the sky. A leopard had spotted them in the trees and was staring up at them. The monkeys got such a shock at the leopard's stare that they had a heart attack and dropped dead at its feet.

African monkeys

Secretary bird

Our safari guide,
Joseph

Thursday

I've just seen the craziest bird: the secretary bird. It has long feathers sticking out of its head and stilt-like legs. I watched it run along the ground after a snake.

Then it killed the snake by kicking it with its long legs. The secretary bird eats highly poisonous snakes, as well as lizards, insects and **rodents**.

Friday

The sound of howls and spooky laughs sent a shiver down my spine this evening. Not too far away, I saw a group of spotted, or laughing, hyenas tearing at the flesh of an antelope and crushing and eating its bones.

Hyenas have very powerful jaws and are fierce predators of large animals, including buffalo and zebras. They are also super **scavengers**. They eat the dead flesh of animals killed by other predators.

A yawning hippo

A hyena

Hippos are vegetarians, but our guide told us they kill more people than any other animal in Africa.

Daily spot

Lions need to kill about 30 animals a year to stay alive. A lion could kill and eat seven antelopes, four zebras, 17 gazelles and two buffalo in a year.

Did you know?

When an antelope is caught, the rest of the herd often gathers in a circle, wide-eyed, to watch the show. Its predator then gets a second chance to make a killing.

Flying high

A food chain at night.

THE Great Horned Owl gets its name from the little tufts of feathers or "horns" on its ears.

This owl is a **nocturnal** (say *nok-ter-nul*) meat-eater. It silently swoops down for the kill, then swallows its prey whole—flesh, bones and all.

The Great Horned Owl gets its food fuel by eating a wide variety of living animals including rodents, small dogs, birds, skunks, reptiles and fish. It also eats other owls!

It has some nifty night-time tactics for nabbing the tasty creatures that it dines on each night.

To spot its prey, it can turn its head around to face backwards. It can also turn its head upside down.

Its soft feathers muffle the sound of its flight, letting it strike without warning. Then it grabs and kills its prey with its powerful sharp claws.

Did you know?

The woodpecker has a clever way to stop getting headaches. It has a special cushion between its beak and the rest of its skull. Also, muscles on its beak contract to soften the blow.

Lying low

Snakes alive—don't eat me, I'm already dead.

THe European grass snake has a sneaky way of escaping being food fuel for a another predator in the food chain—it acts dead when it is attacked.

When this snake senses danger, it goes limp. It rolls over so part of its belly is in the air. It rolls its eyes back in its head, opens its mouth and hangs out its tongue.

It lets out a stinking liquid that smells of rotting flesh, so it even smells dead—really dead, then a passing predator isn't tempted to take a bite.

But it can get carried away with its play-acting. If it gets turned over, it quickly rolls back into its "dead position", in an effort to prove that it is not alive.

This snake feeds on small creatures such as frogs, toads, mice and birds. It has a simple way of catching and eating them. It bites its prey and then swallows it whole, while it is still alive.

The News Cartoon

Action in the ocean

Shocking rays and shooting scallops—food chains under the sea.

ELECTRIC RAYS lie on the ocean bottom, waiting to zap passing prey. There are special muscles on each side of the head of these flat fish. These muscles act like batteries and produce electricity. The rays stun their prey and then eat them.

The torpedo ray is a real shocker. It uses 220 volts of electricity to stun its victims. This jolt makes its victims jerk and twitch like crazy.

The scallop is a smart **shellfish**. It can make a speedy escape so it doesn't get grabbed by a crab or snared by seastars.

The scallop shoots off the ocean floor with a jet of water that it spurts out of its shell. It uses this jet stream to quickly swim to safety.

When scallops feel like a tasty meal, one of their favourite foods is plankton.

Other **molluscs** aren't so speedy. When they need to hide, they bury themselves in the sand or stick onto rocks.

Plankton are tiny plants and animals that float in the ocean.

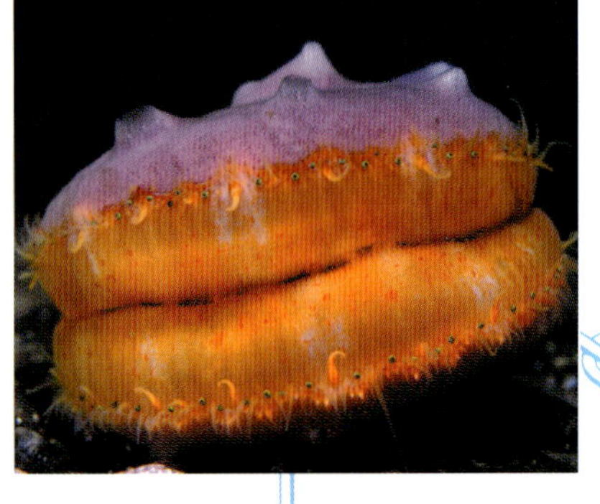

Scallops snack on plankton.

Did you know?

The great white shark is at the top of the food chain under the sea. But it can't see when its eyes are cold. It warms up its peepers by shifting heat from the muscles of its body into its eyes. Then it can see and catch even the fastest fish in the sea.

Scallops are eaten by seastars, crabs and people, too.

Hide and seek

Fighting fish gobbled up by father but lives to tell the tale.
Imagine… if she could tell her side of the story!

SHARON IS A young Siamese fighting fish. She told *The News* about her father's strange ways.

"My dad's a bit of a strange fish. He fights a lot with the men in his family. But he's a really caring dad," said Sharon. "He looks after all 120 of us all by himself."

"He has his fins full, just making sure we stay out of trouble in the weedy waters where we live. He even has a special alarm system to warn us about danger."

"When a bigger fish looks like it wants to make a meal of us, Dad shakes his fins to make ripples in the water."

"This tells us that we are in danger and we should swim over to him. When we get close to him, he scoops us into his mouth."

"We stay hidden in there until the coast is clear. Then he spits us out again. It's a really fun game."

Animal of the day

SIAMESE FIGHTING FISH—a small freshwater fish from Thailand (which used to be called Siam) and Malaysia. It has long fins and is brightly coloured—usually red, green, yellow or blue.

Underwater war

Platypus uses secret weapon to find food.

THE PLATYPUS eats half its weight in food each day. Worms, crabs and shrimp go into hiding to avoid becoming its food fuel.

A platypus has a furry body, webbed feet, a bill like a duck and a tail like a beaver. It is also a monotreme (say *mon-a-tream*)—a mammal that lays eggs.

When the platypus is hunting for food, it shuts its eyes. It also closes off its other senses so it can't hear or smell. But to find its food, it has an amazing secret weapon. It has an electrical sensor in its bill.

The platypus sweeps its bill from side to side. As worms, insects, crabs and shrimp move around in the water, they give off electricity. The sensor in the bill of the platypus can pick up these tiny electrical signals.

So, as the tiny creatures hurry to hide, the platypus is able to find them. In this way, the platypus can catch and eat all the food it needs.

Shrimp are eaten by people as well as by the platypus. Shrimp are not red until they are cooked and ready for eating. In the ocean, the naturally dull colours of the shrimp help them to hide from predators.

The platypus is only found in rivers and lakes in eastern Australia. The platypus is such an odd-looking animal that the first time a specimen was shown in Europe, it was thought to be a fake. Experts thought it was pieces of different animals sewn together!

The wolf and the heron

Ever heard the expression—**STOP** while you're *ahead*?

A once asked a heron to use her long bill to get a bone that was stuck in his throat. He promised her that he would give her a reward.

The heron got the bone out easily. When she asked for the reward, the wolf just laughed.

"You stuck your head in my mouth and you didn't get it bitten off. That's a reward."

The heron wasn't happy. She saw some birds pecking inside the mouth of an alligator. They were cleaning its teeth by eating the food stuck there. It gave her an idea.

"Let me clean your teeth, like those birds are for that alligator", she said to the wolf. Her plan was to jab the wolf's throat with her sharp bill.

The wolf wasn't fooled. He pretended to agree and opened his mouth. When the heron leaned in to his open mouth, he swallowed her in one gulp.

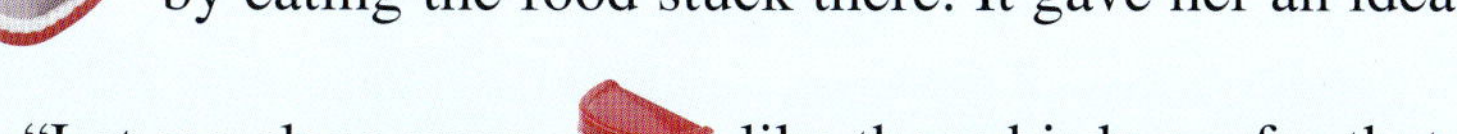

wolf

heron

mouth

teeth

alligator

Pack attack

At the top of the food chain, wolves and whales win out in the cold.

THERE'S POWER in numbers at the **poles**. Some polar animals hunt in groups.

The killer whale in the icy water and the gray wolf on land, both attack in packs. This gives these fierce predators a winning edge. In packs, they can kill animals much larger than themselves.

Gray wolves hunt in packs of up to 20 animals. They prey on large mammals like deer and moose.

The wolves travel in single file, trotting for hours to tire out their prey. When a kill is made by the pack, the wolves all try to tear at the meat at the same time, and they eat it in a rowdy, growling, scramble.

Killer whales are sometimes called the "wolves of the sea". They are very fast swimmers with large, sharp teeth. They hunt together in packs of 6 to 40 whales. These packs are called pods.

The pod works together to surround and force the prey into a small area before attacking.

Killer whales use different tactics when hunting very large prey, including other whales. The pod will usually attack from several angles. They even hunt the enormous **baleen** (say *bay-lean*) **whale**. The killer whales swim into this giant's mouth and bite off a piece of its tongue.

Wolves eat caribou (reindeer), musk-oxen, bison, Dall sheep, elk and mountain goats.

Killer whales eat fish, seals, sea lions, whales and other marine mammals.

Word of the day

POD—a small herd of seals or whales.

Whale food wanted

Finding enough food fuel for the world's largest animal.

BLUE WHALES are the biggest creatures ever to have lived on the Earth. They can be up to 33 metres (108 ft) long. They're even bigger than dinosaurs.

But blue whales are now **endangered**. There are only about 2,000 left in Antarctica.

For years, people hunted blue whales for their oil and meat. Humans were the predators of the blue whales. In one season, 30,000 were killed. Blue whales became so rare that laws were passed to protect them.

These gentle giants eat krill—tiny, bright red, shrimp-like animals. One whale can devour 40 million of these micro meals in a day.

When the blue whales were over-hunted and their numbers reduced, there was more krill for seals and penguins to eat. Because there was more food, the number of seals and penguins increased.

Now, with so many of these other animals eating krill, there is not enough food for the blue whales.

The food chain balance has been changed because people killed the blue whales. Now there are too many seals and penguins sharing the whales' food supply.

This krill is shown at its actual size. It's no wonder the huge blue whale has to eat so many of them.

Daily spot

Blue whales talk to each other over huge distances. They use low-frequency moans which can travel thousands of kilometres (or miles) through the water. Close up, these moans sound as loud as a jumbo jet taking off.

Animal giggles

Food fun.

Bold Jenny and the bear

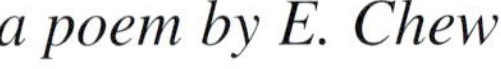

Jenny met a big brown bear,
But she didn't even care.
The bear glared,
But Jenny wasn't scared.
The bear looked cruel,
But she kept her cool.
The bear was in a flurry,
But Jenny didn't worry.
The bear got ready to maul her,
Jenny rose up taller.
Then wham, punch, crunch!
She had him for lunch.

a poem by E. Chew

JOKES

Q. What happens if you feed cocoa to tadpoles?

A. *You get chocolate frogs.*

Q. What do you call a pig eating in a restaurant?

A. *A pig out.*

Q. Where do monkeys cook their toast?

A. *Under the gorilla.*

Q. What's a hedgehog's favourite food?

A. *Prickled onions.*

Q. What does a frog drink when he's on a diet?

A. *Diet croak.*

Follow the food chain

Make your own and watch it at work.

THERE ARE millions of food chains in the world. Here's one that starts in the garden. See if you can work out where it ends.

You will need:

- a stem from a rosebush
- an aphid—a small insect that feeds on plant juices
- a ladybird—a beetle that eats aphids
- one small and one large jar
- cotton wool
- a magnifying glass

What to do:

1 Find a rosebush that has aphids on the leaves or stems. Ask an adult to help you carefully pick a stem. Take care not to let the aphids drop off.
2 Fill the small jar with water and put the stem in it.
3 Put cotton wool around the top edge of the jar to hold the stem in place.
4 Place the small jar inside the larger jar.
5 Watch the aphid through the magnifying glass.
6 Notice how it sucks the sap from the plant with its thin, pointed mouth.
7 Put a ladybird in the jar. See it catch and eat the aphid.

Where to from here?

You can draw a picture of what happens to the food chain from here. Start from the rosebush and draw the food chain so far.

Ladybirds are eaten by birds. What do you think might eat the birds? What if the bird that eats the ladybird is a chicken? What or who might eat the chicken?

Where do you think the food chain might end? Have fun finding out!

Editorial

SOME people think that taking one link out of the food chain doesn't make much of a difference. But every living thing relies on another to stay alive. All links in the food chain are important.

For example, if there are not enough meat-eaters to eat the plant-eaters, then all the plants get eaten.

In India, the bullfrogs from the rice fields were sold to countries in Europe where people like to eat frogs legs. Without the bullfrogs, the insects they used to eat were able to increase until they were out of control. The insects ate the rice crops so there was not enough rice to feed the people.

When humans disturb food chains, they upset the balance of nature. They harm themselves as well as the plants and animals of the world.

At *The News*, we believe that by understanding how food chains work, we can make sure that we do not destroy important links.

Letters
to the editor

Dear Editor,
Yesterday I saw a cat catch and kill a bird. This made me upset and angry. I think cats should wear bells to let birds know that there's a cat nearby.

From BILL, THE BIRD WATCHER

Dear Editor,
People should stop killing African wild dogs. Some people think these dogs are cruel and savage. But these dogs do an important job. They kill sick and weak animals. They also keep the numbers in the herd down so there is enough food to go around.

From VICKY, THE DOG LOVER

If your pet bird got out of its cage, your pet cat would eat it. You would be very sad, but that's the way it's meant to be. It's the food chain.

Zoe, the zoologist,

answers your questions

Dear Zoe,
I was told that jaguars are man-eaters. Is this true?

From MOLLY TAN

ZOE: No, this is not true. Jaguars are not man-eaters like the other big cats. Jaguars have killed humans, but only when people hunt, taunt or wound them. Jaguars eat a wide range of foods including wild pigs, caimans, anacondas, capybaras, birds, turtles and fish.

Dear Zoe,
Do snakes only eat meat?

From TARA BRYSON

ZOE: Most snakes do eat meat. They eat frogs, mice, termites, lizards, other snakes and birds. Some large snakes can kill and eat big animals like deer, pigs, caimans and capybaras.

One species of snake only eats the eggs of birds, lizards and other snakes.

Dear Zoe,
I thought the rhinoceros was a savage animal. But my friend says it is actually a harmless vegetarian. Who is right?

From BILL BROWN

ZOE: Your friend is right. The rhinoceros is a herbivore who eats only leaves, grasses and twigs. Its two big horns are used for protection, not hunting. They are made of fibres that are similar to matted hair, not hard bone.

Humans have killed large numbers of rhinoceroses for their horns. These are used to make many things such as medicine and dagger handles. Rhinoceroses are now an endangered species.

Jaguars may not be man-eaters, but it's probably not a good idea to cuddle up to one.

Animals of the day

ANACONDA—one of the largest snakes in the world. It spends most of its life in murky, fresh water. But it can also climb trees. It kills its prey by strangling it.

CAPYBARA—a large rodent that looks like a giant guinea pig. It lives in forests, near water, in South America. It is an excellent swimmer and can run like a horse.

Beware the rhinoceros!

Further resources

BOOKS

Most books on wildlife will tell you a lot about what these creatures eat and what eats them. There are also books specifically about predators and prey. Look in your local library or at school for these titles:

Anita Ganeri *The Hunt for Food* Belitha Press

Peter Riley *Straightforward Science: Food Chains* Watts

David Taylor *Animal Assassins* Boxtree

Jenny Wood *Animals in Action: Animal Hunters* Hamlyn

MAGAZINES

There are lots of wildlife magazines, such as *National Geographic*, which have terrific photographs and stories. There are also magazines especially for children. These include *Explore*, *Comet* and *Go-Magz*. Look for them in your local library and at newsagents and bookshops.

NEWSPAPERS

Newspapers often have wildlife stories and reports on issues such as the protection of endangered species. Local papers also have articles about bird and animal life in the local community.

TELEVISION & VIDEO

Keep your eye on the television guide for nature and wildlife programs. There are also some fantastic videos including *Supernatural*, *The Unseen Powers of Animals* and *The Greatest Wildlife Show on Earth*.

INTERNET

There are a huge number of web sites that have information about animals, birds, fish and reptiles. Try the following addresses and link to the sites that you are interested in:
www.nationalgeographic.com/kids
www.animaldiversity.ummz.umich.edu
www.britannica.com

The Pitcher plant is a carnivorous plant. Insects are attracted to the pitcher, fall in, drown, dissolve and the food is absorbed by the plant.

Fact of the day

Some rainforests have flying frogs. These little critters jump out of trees to escape from predators. They have large webbed feet and hands that act like parachutes as they sail through the air.

Glossary

BALEEN WHALE—(say *bay-leen*) a type of whale with bony material hanging in its mouth to gather the small animals it eats.

CARNIVORE—(say *car-niv-or*) an animal who eats other animals.

ENDANGERED—an animal whose numbers have been reduced to the point of nearly disappearing.

FOOD CHAIN—the way that plants and animals are linked together by their feeding habits.

HERBIVORE—(say *her-biv-or*) an animal who feeds on plants.

MOLLUSC—a soft-bodied sea animal in a hard shell.

NOCTURNAL—(say *nok-ter-nul*) active during the night.

OMNIVORE—(say *omm-niv-or*) an animal who feeds on both animals and plants.

POLES—the two ends of the Earth.

PREDATOR—(say *pred-ah-ter*) an animal who kills and eats other animals.

PREY— (say *pray*) an animal who is killed and eaten by another animal.

RODENT—a rat-like animal that has teeth for gnawing or nibbling.

SCAVENGERS—animals that eat the dead flesh of other animals.

SHELLFISH—an animal with a shell, such as a prawn, a scallop or an oyster, that lives in water.

SPECIES—a specific kind of plant or animal that can mate and produce young.

Did you know?

The basilisk (say *bas-ill-isk*) lizard uses flaps of skin on its back legs to run on water. These flaps trap air and stop it from sinking.

Tasmanian Devils are scavengers. This means they are near the top of the food chain. They eat leftovers—things that other animals have killed.

Index

Mushrooms are at the end of the food chain. Mushrooms live on dead plants and trees. They are the last link in some food chains, then they are eaten and the cycle begins again.